Masters of Life and Universe - Inspirations for Pro Self-Replicating Technology Movement

"A new heaven and a new earth"

"Then I saw a new heaven and a new earth, for the present earth and sky had disappeared. There was also the sea.
 And I saw the holy city - the new Jerusalem, coming down out of heaven from God. She looked great - like a bride who beautifully dressed for her husband.
Then I heard a loud voice coming from the throne: "From now on, God will dwell among the people. They will be his people, and he Shall be Their God
 and He will wipe Their eyes all tears. There will be no more death, sorrow, crying or pain. All this is gone along with the old world. "
 Then the one who sits on the throne said, "Behold, I make all things new." Then he turned to me:- Save it, because These words are trustworthy and true." (Revelation. Jan 21)

Table of contents

1. Lunar workshop of the world and space
2. New Adam and Eve
3. Conscious citizenship in universe environment
4. Creations power of God
5. Self duplicating plants
6. Body cell-like replicating productions
7. Real automation investment
8. Just for fun replicating investment
9. Taming horses and… resources
10. Paradigm of subduing forces of nature and their replacement
11. Peoples free zones – small "moon" zones for everybody
12. Amanhattan
13. We are only little step from being masters of whole life and universe

This is an introduction, draft, textbook. This "brain storm" hand-pocketbook is to widen up horizons on our existence, but not quite in a philosophical way. You may be more inspired in innovations proposals that deals with climate towards fundamentals for modern civilization. This includes everyone and the entire environment including his surrounding space as well. In this book we have talked about interplay of economic, technological, ecological, and crucial factors of our immediate environment that interfere in our biological existence.

Generally the introduction book is about my -still by constructing- huge project Moonnow.

MOONONOW is about creating peaceful global and local organizational and technological conditions for hyper active participation of people and technology in local, global, cosmic transformations in full sense of the word.

What is the purpose of the MOONNOW project?

Are super cheap passenger flights to the moon and beyond? Yes, but it is just an epiphany - a string that draws attention to proper teapot and development path - a shift from evolutionary social economic and political stagnation to goals above all to move to a higher level in development of plans for visions of constructions, that completely will shape us and our civilization of our surroundings technology on a macro and micro scale of our cosmos in literally and immediate mode !!!

We have full potential for development of talents, opportunities to overstep impasse of primitive social political economic relations. With help of appropriate organizational and technological changes.

The courageous decisions in building the foundations of technological hyper-technological as well as engaging in this strengthening of proper socio-economic bases give chance for true developmental change in the broadest sense of the word.

It is not about replacing theory of evolution with the theory of creativity, but by leveraging achievements of evolution of technological development into further irrepressible development, which can be characterized by creativity and hard tangible evidence of the existence of God as the omnipotent creator, creator, proving creative theory. We can create things, creatures, and ... worlds for ... a flick!

Proposed technological solutions, which will describe in details later in this mega construction, have full grounds for creative evolution in creation of cities, whole systems of

structures of great or small scale, as at a touch of an enchanted wand, confirming tangibly theories of the divine Creation of the earth and living beings on the earth !!
There will be no time wasted and we should not waste our chances and talents to make the world better, because creation and evolution together give tools, consent and ordinance to make our world more human - divine - merciful and just for each without artificial borders of politics or customs!

So it's time to start this moon's odyssey.

1. Lunar workshop of the world and space

Moon as most effective sources of cosmos exploration – for now. Endless source's energy and materials for building new bases, factories, spaceships and towns on the moon, our orbit and for other places in cosmos.
Let start for using the moon opportunity to speed our enterprises to goal our -not only- space dreams.

Let's try to talk about futurology and it is nearest in today time as possible - first man historic landing on the moon nearly half a century ago already!

Moon perpetual motion investments as the dominant global economic and social locomotive.

It is a program of technological nature of the dominant influence of the expansion of space as an engine of economic growth in the global structural changing.

It proposes to use the moon as an industrial park to multiply the effectiveness of the cosmic expansion, which would become the locomotive of the economy, technological change in the world.

Contrary to appearances, the cost of implementing this program will be negligible to global effects and it is for a very short period of time -10 years for at least building first towns and hotels on the moon.

Using the support of foundations, space agencies to send probes with mini robots, printers 3D to build factories next robots, extracting rich minerals and further creating of the next growing processing for increasing space investment on the moon and further, all with appropriately processed the vast infinite solar energy on the moon.

Moon as a testing ground for new technologies, and especially here hinted technology, 3D printers, and robots for building up the next robots and on that next factories and robots with multiplications effects and structures with ever greater speed. I mean here also in this case that is not a long time by the speed of light! So the formation of cities or giant spacecraft in such way like the formation of a three-dimensional image in a few seconds!

And here by the formatting of the moon cities, we could look for like new theory ---"constructions speed multiplexing" –, which moves away as one time theories of Newton and now theories of Einstein.

-theories Einstein used to nuclear energy - let's call it the energy of Satan. But the like theory of Speed multiplexing construction could be named as an energy of God, where the universe becomes for us as close as small backyard and it would be consequences in macro cosmos and also in micro cosmos matters – medicine. But to come to that I need help - as it was for example by electric bulb construction Edison used help of teams to work out it.

All these moon constructions could be effectively guided remotely from the Earth ground due to only one second light distance to the moon. Anyone could take part in it with the appropriate mobile applications - as well used also mobile games. Of course, in time the moon investments would be fairly autonomous - full automation – in domino effect way as well as exponentially developing space economy, space industry, where the weight of products, eg. Space stations, probes were sent to continuously in the cosmos and back into orbit around the Earth or Earth too. Space shuttles - also built on the moon flying between the moon and the earth will facilitate the circulation.

Not forgotten here for moons populations for production bases, research, and hotels, etc. Before you can tentatively establish such a base, eg. We can do it first base for rats.

The main cost of the initial project to raise the barriers of gravity of the earth to orbit the Earth, because the rest of the transmission of the products of the industrial park on the moon back and forth is a low-cost flight in the vacuum space. Most of the products will be formed on the moon. Only the transportation of people and other unattainable necessary means missing the moon will be most entered in those transport costs.

I pointed out here factor in social, economic, political matters. With such progressive expansion in the planned development of the moon-economy moon- space economy- in the geometric domino effect of perpetual motion, the role of human will be more likeable voluntary participation to work – not normally work in these projects beyond the earth and ultimately all projects also on the earth, which completely rearranges the system of economic , social, political relations in personal life and work and internationally, and this in a positive sense.

Organizational help for encouraging schools, universities, foundations, businesses locally and internationally to support with ideas, foundations, companies also countries such like big tourism investor Dubai for the moon undertaking.
The enterprise can start with programs, games, mobile applications, after some period also of sending moon probe, which would start described lunar spacecraft domino effect. And this all could start from foundation a special website for media to support this project ideas advertising.

. We all need a support of encouraging us to reach not only the Moon, planets or stars but anything that we aware or not aware, that could improve ours private or public life.
This project is not only about "vanity" cosmos hobby tourist matters but also to gain by the way matters of ours existence in an economy, political, technology matters – ecology, medicine –developing also small robots to fight against poverty, mortality diseases, … mortality for everybody!

I need peoples movement by encouraging for support by state and private investments, foundations, sponsorships.

This project can start with minimum budget almost everywhere, where technology is enough capable to build Moon probes and self-producing robots(in like Lego system)– for example USA, Japan, India, Russia, UK, Poland and so on also big private firms and foundations, universities.

2. New Adam and Eve

Project MOONNOW is about building (with our hands and our applications) new mankind civilizations of robots on The Moon, that help us to positively rebuild our world. Generally, it is about to start the robots domino effect with ever faster speed. What will build over the moon in a few years. Not to mention the spacecraft and technological changes (social, too).
It is like first people started their civilizations on Earth using the earth resources, so the same will be with robotics moon, but with much greater speed.
What I need is technology, organizationally, eventually economical support + extra ideas in the moon robotic revolutions. We need at the start one simple moon probe and simply firsts robots to start it. The first simply robots using almost only moon minerals and solar energy will build another robot. They will build faster, more efficiently, their successors factories and cities for us. What is your opinion about the future of aerospace and hence astronomy - but not the very distant but the closest to us? So far our development of aerospace and other technologies for the design lies in its infancy - it's just pure manufacturer - in relation

to our informatics capabilities. Of course, speaking here, in general, is the state of the global situations.

I propose output from the inability technology.

So everything should start from "the beginning" but already at quite another level.

This is how Adam and Eve began to populate and civilized our Earth. But here the Adam and Eve may be the most simple robots, which will be able to construct next robots and other structures of raw materials (mining, smelting rich resources of the moon)and super rich solar energy, in this case on the moon. And we'll be the God directing the lunar beginnings of civilizations, with (breath taking) consequences for the whole of our civilization, in a quite short time. Maybe would you have made some sort of brick is the suggested project?

At the beginning of this hard work will be arduous wet behind the ears. Just as it was with the development of computer science with primitive systems of zero and one.

It is about invasions of the moon, that can budge our chances of further entirely development.

This project includes changes in the position of a citizens, whether they would like to take part in this or not. Not only they will come here in open participation with help of phone applications but also economically support.

I am interested in sharing my other views and suggestions with institutions of different views. I am interested in deepening the theme with every person who would be interested in experiments, and help in further strong growth on this topic. Maybe it would be possible to have a closer look together for exchanging materials and experiments on this topic.

Political social media as chance for wakeup of ethical economic awareness for really sustainable development of civilizations improvement

3. Conscious citizenship in universe environment

The "moon" construction - first on earth -should start from organizationally matters. First at all it is about extra ordinary situation of extraordinary construction foundations for waking up, stimulating - activate- awareness and instinct of (better) life of scientists, technologists, citizens.

Create organizational foundations for civilization jump, that more precisely could fit our individual and collective needs, dreams and of course potentially enormous opportunities. And all this with exclusion of religious and national artificially barriers.

A global "environmentally, cosmic-friendly" foundation based on a voluntary partnership of scientists, technologists, doctors and all citizens - something like the Doctors Without Borders - would become a model of public relations for now and tomorrow, here on earth and in space. It would take -in next future- over from today's systems (governments, economy), to a smooth degree, all most important public, economic, developing functions.

I cannot impose this or other projects gave the whole world, and every citizen individually. This project includes changes about position of a citizen, which he would like to take part in this or what possible public enterprise . And not only comes here in the open participation of the help of phone applications but mutual support economically.
Those interested more on this issue more accurately, they can trace my attention to this part below.

Developing of citizens involvement for more equitable environment resources

Public (economically) toll for conscious shaping own
entire environment space and whole future.

Awareness of mutually needs of both sides of consumer citizens and of other sides capitalists. We should find out language/right tools to work out at right level of communications between both sides. I propose outline and suggestions for more democratic style of our local and global politic considering social, economic and civic poverty. Climate and other global, cosmic issues that are very important for better development of world and any citizen's needs separately.

Alternative for the bankruptcy of the capitalist model could be implementation of environment global tax system, that would award new pro environment, pro development global positively enterprises (new investments , new technology and so on). All in assumption of international conferences, organizations like UN, EU, states multi agreements, but also take in big consideration of all citizens (vote) voices part without their participation, all afforded enterprises will be finished with wrong effects. Governance of only economical, political technocratic group in today's world is not enough. We all should use new technologies and take chances of making our world out of stagnancy social and technological common good. We cannot speak about ethics of relationships without right developed democratically, economically, awareness of mutual needs of both sides of consumer citizens and of "another" sides "capitalists" of course in bigger or smaller variations. Every person represents needs and behavior of "another" sides. Not economics animal language but a progressive human language is right way towards more ethical relationships of our global present society for new communications renaissance after 600 years old of printing renaissance of communication.

But everything would fail if citizens wouldn't have chance to directly take part in the global tax decisions. Well elaborating dialog (not monologue of economics' technocrats politicians) language can be key to improve effectiveness of political compromise, mutual communication of both sides i.e. between citizens and political and economic technocrats. Citizenship aspirations, developing trust for public enterprises can be supported by more democratically, serviceable, practical, equitable dialog. I offer outline of an equitable tool that can rouse and support people's citizenship aspirations and take responsibility for the limited environment that really needs development. I propose that administration, statesmen should prepare to take permanent survey, opinion (votes) collector system to make more reliable, precisely make adequate tax revenue spending. Tax spending opinion polling votes inquiry

system where people taxpayer could decide where in state, public enterprise issues should spend part of the taxes. For example - about 10 x 100$ a piece or so to say 10 x 100$ votes.
The polling(voting) right of every person would depend on amount taxpaying of every person. But could be appointed minimum the tax vote right for every person even for those who don't pay tax.
I propose that it could be used for banking account system where citizens could show their own opinion about their own tax spending(by using the x 100$ votes) for public enterprises through their own special tax account by using system of home banking, post gsm, mobile phone and so on. Also by using the same form of the inquiry could be made annually permanently. General referendum about growing or decreasing taxes, weighted average of the tax votes (about 10% less or more). It should be immanent part of the whole proposal. So that people could decide which direction (part of their own earth-share, universe-share money) should be moved, not imposed for things- even cosmonautics- that maybe people would not like to participate, where would it be more needed (outside or inside public matters - macroeconomic automatic stabilizer) to make improvements and investment accordingly. All low regulation tools are almost ready. We don't need to make big changes to the laws because it is already written that people have right to show their own views and that authorities have right to use different tools to collect opinions from voters.
What we need is to elaborate the new "post Gutenberg" language, the enquiry bond vote- forms ("tax shares" system) to properly elaborate to the departments to take right responsibility for attendance of the all year tax spending votes enquiries.
Taxpayers, citizens are in reality more likely to be shareholders,

investors of their public limited environment resources.

Of course not all people takes part in typical shareholder market, same is the case with normal elections to parliament, but that gives no right to cancel such institutions. Similarly we cannot deprive people's chance, opportunities, right to express their own opinion about their own (public) money, share and responsibilities. We can elaborate more accessible (not complicated donation tax system) public spending enquiry. Proposed tax opinion enquiry (language-barometer) can rouse support of citizenship aspirations; fight against social and economical poverty. The opinion barometer (macroeconomic automatic stabilizer) like shares in free market would be used for reliably collecting opinions on public enterprises as anonymous enquiry bond-form (like check, banknotes) to point out opinions votes of the taxpayers (With help of attached guidelines and schools pro civic classes). The taxpayer's ("tax shareholders") in these forms could decide where, which department of public matters should get their tax money support. In indicated locals, offices (maybe by post too) each person could show where should part of his paid tax go (e.g. 20%) - It would be minimum allowance for persons, who can't pay tax. Earlier elected in normal elections statesmen would be responsible for taking proper consideration of the tax spending referendum enquiry (it can be used as WEIGHTED AVERAGE by sharing the all enquiry votes.
The lag between fiscal decision and effects of the decisions would be smaller. We should properly and steadily elaborate precise democratic dialog tool for more reliable economical automatic powerful – for enormous enterprises like intensive automatically moon exploration- stabilizer that could be used in tax system by

taxpayer, citizens, politicians, and statesmen.

Tax –public money (money is usually treated as equitable, precise message about needs and opportunities) we have to understand like mortgage or fee of widely taking in consideration environment resources. Citizens will have right of taking direct influential share -decisions and profits- exchanging (public money) messages about need, opportunities that can push development in needful direction.

Technocrats cannot always know where to spend public money. The politicians need more proper reliable opinions about needs of ordinary shareholders of state of the technocrat politician, but above all state of the people. Citizens-consumers need such tool (psychological toy) to be more firm to gain foothold wind up in public matters that absolutely interfere with so called private market ("free" market matter). I do not want to replace law, that all people should replace statesmen's
work. I want to give the statesmen more effective toll, barometer of people's environment needs and opportunities. So decision of the statesmen would be wiser, connection between politician, statesmen and public, community more objective accordingly to really opportunities and needs everyone. The tool, the opinion barometer (macroeconomic automatic stabilizer) like shares in free market would be used to collect more reliable opinions
to make more firmly, courageous, more approval decisions on public investment, enterprises matters (tax matters) according to needs and opportunities of the taxpayer, citizens, "shareholders" of taxed state environment resources.

All taxpayers, citizens will have absolute right to get a chance to be in CLEAR, FAIR participation sharing - in own taxed (shared) environment. So that definition of tax could have quite a different shade. People will have more trust in so called tax money which is very important in making (economical) public matter more effective (that pushes in right directions effectiveness of revenue –further economics utilities of taxation- of Laffer curve.

All interested from abroad (on strict conditions) to get the public money support on particular public issues could take part in competition to raise "tax money share" through the tax spending enquiry survey. Public investments, space and another enterprises offers can get new support, new sustainable toll to boost entire economy in safety, salutary (for our environment resources) directions. People could directly decide, where – in space- create more sustainably enterprises and workplaces. People can give big argument to support entire economy included any enterprises, wherever it could develop, whether it is outside or inside state budget. People could better appreciate power, service, utility of money and tax money. Of course my proposed system would evolve by new technology, banking system where politicians could be less intertwining by controlling of the system, effects of the tax opinion enquiry would be more automatic (macroeconomic automatic stabilizer) and instant like use of shares, stocks in stock exchange under suitable regulations.

All institutions whether public or private would rival each other to get more chance of taking so tax votes in the tax spending referendum enquiry. They can get further money from state authorities after budget approval. Gratis, local newspapers, booklets, magazines could give support to this enquiry enterprise.

The public enterprise could elaborate readers and editors at least in council taxes matter range.

Also local newspapers for example could be issued with published forms where voter could describe where (local or globally). The enquiry forms like money (check or bond) in free market can improve exchange messages of opportunities and needs down to earth therefore more pro-environmental of all people and citizens. Today social media like Twitter , Facebook could be give big social and pro civic media support for it. Statesmen (partly) like staff of banks or Stock Exchange would take proper attention of the taxpayer "bonds opinions". The enquiry forms, the new money, the opinions enquiry forms can positively influence global, psychological , economical, political matters, can really help in fair development of our entire environment and life and fulfilling our hopes. Every person, citizen, consumer simultaneously could have new powerful, helpful public toll for responsible forcibly shape, harmonize their own entire environment and future by also powerful enterprises- showed on Moonnow.

4. Creations power of God

And now I will try to explain some possible core of the anticipated futurology irreversible changes in methods of use of construction technology, but in my outline the changes could, they have to be - must take place immediately.

I pushed them all possible ways. Witnesses can take part in this project, using the transfer my or their proposals in this topic further.

Moon as a testing ground, it may be a giant, autonomous science, and industrial park.

To conduct this park we will need a good structure telecommunications – "GPS" - between the Earth and the Moon, and as I wrote the beginnings of a remote-controlled and automatic robotics constructions.

As I have written abilities and telecommunications now and in the near future, are at least very promising. We lack harnessing to our brain telecommunications, agile hands, able to use the potential of ICT.

May these hands be autonomous robots fully cooperating with ICT systems, which will continue, of course, subject to the power of the people.

How can these robots operate?

Evolution or I would here call mandatory revolution consisted of robots to the relatively simple principle - one hand washes. The robots could autonomously produce his own incubators. Of course, by our proposals and disposal of ICT, so that they would serve as our hands to build the big and small things, but with the great difference that they will do so in an infinite and ever shorter time and in the infinitely any number and size.

These robots are automated as our children - taught, educated, prepared, refined to them to be self-sufficient and at the same time help to further our and their development.

I know it will be hard to wean us from traditional manufacturing approach to the production of these impersonal autonomous robots parks. This can cause mental, organizational discomforting situations. - Moreover, as far as I know – such full automatics factories already exist.

Besides, the robots will produce products that the market is not, so there should be competition, but so contrary impulse for further projects, which will need an infinite number of people - coordinating ICT - automatic guidance expansionally giant space programs and others.

We have to start intensive work out - the same system for constructing self-repairing robots and construction. 3 D printers are a great example and the beginning of this process.

The beginnings as always can be difficult.

One might say that this project is crazy. If this project would be crazy projects sending people to the moon or Mars in the present - lack of facilities of basis - call idiocy.

I believe that with the right – very expansively intensively automatic (without directly taking part of people) - automatic preparation of bases between the moon and the earth and other planets is a chance of effectively populate the moon and so on. So construction -for example- of tourist resorts on the moon could be available for about 10 years. Finally, for example, my wife – on behalf birthday could get a gift from me , in the form of a 2 week trip and stay with me on the moon in 2025(that all in a price of today's trips to Antarctica).

But art is there to fly comfortably, safely, cheaply and ecologically. Until they met these conditions neither I nor my wife, nor any normal, responsible person goes the today way. So hyper intensive automatic preparations give guarantee for successfully moon, space explorations.

I propose something different - incubators automatic robots will nucleus proper feedback ICT and automation and after all, human will coordinate these processes.

Someone will say after we produce robots - so but for the most part by very primitive manufacture ways so at already it at start eliminates any efficiency move forward our civilization even one iota.

It must be whole teams of robots, their incubators and factories that mutual feedback and ICT support each other will drive right effective performance.

I'm not interested literal stupid hammer -welding space bases or policy makers, for whom the status quo in the methods of gaining space methods from before 50 or 100 years is the only paradigm of progress.

Tedious folding nucleus of the robot empire will disproportionately benefit in relation to the already literally manual gain space - huge multi-gear multi-engine itself, extending the up increasingly powerful pace spaceships - built on the moon and city on planets, processing structure, acquire, protect our land - ecology, our life figuratively and literally - armies of micro and nano robots in our body for remote and autonomous repair and maintenance to improve all parts of our body.

I earlier hinted at the power of God, which is what the energy of creation in the blink of an eye - this is the above-described energy - which sort of proves the theory of the creation of the world or the universe by God or Big Bang. In this case, it is within our hands. We kids (and its hinterland of thought, potentially technological) of God to meet the outstretched hand of God, which is in the - near case- the Moon, we can if we believe in our abilities and capabilities, create, create the world, literally the entire universe by own dreams and wishes, and it would be possible as soon as possible in not so stretched time!

Someone can tell me a quack, Futurological Andersen or cosmic Harry Potter, crazy Einstein, perhaps the awaited messiah end of the old world – I think not but just the prodigal son of the Father. Or just the one who brings new hope for the good news - the new gospel that shows new chances of using our talents and possibilities and ways, new methods to fulfill these good news – cosmic evangelism.

Now let's use this opportunity for us, for our loved ones - as described above indicated the hand of God - the Moon.

The key to the success of the project MOONNOW is an autonomous system mechanism large or small or very small, which repair by themselves, develop, multiply - that is, and how as a printer 3 D - BUT !!!! this printer must itself reproduce, repair, develop, improve, to grow or reduce to. It is like in a computer system to zero one nature, which we use to infinitely many records or calculations.

Yes, incubators robots that will continue to grow infinity degree in any direction depending on your needs - more or less intensive control of people.

These robots or incubators or other more or less closed systems will serve us for investment on an unprecedented scale will be only limiting by amount - almost endless –material and energy- including atoms energy- in space.

This kind of closed systems is like reproducing human cells, the reproducing virus – a cell for cloning - adequately controlled, producing interesting part of us and other systems at the macro and micro. These are our technological children, which can be driven each other to increase by infinite atomic energy, the energy of the stars.

If we are children of God, they are the autonomous closed systems are his grandchildren.

If God or evolution gave us a brain, its development was addicted to each other from our hands and their dexterity. That our ability to effectively actions are and will not only depend on the telecom computing capabilities but also efficient multiplying body cell - described above autonomous mechanisms - more efficient than stumps current technological capabilities.

These grandchildren it will be those supermen (not necessarily like us) - the macro and micro, which will be able to casually conduct mobile applications or even games.

We need support like special zone scientific- technology project Manhattan , but on open for all citizens and global and peacefully scale , with the support for example mobile phone applications and other ,mobile services for consultation, programming among others : energy supply, building material or location and directions of development of the autonomous structures, which can be e.g. Towns on the moon, spaceships or modified mini robots - incubators in human body.

5. Self duplicating plants

The moon is the best space out port under the sun, which may become the largest economy Space (a giant industrial park with two times of area of Russia) under the sun.
It is like a big mine, the great spaceport, any bureaucratic only great freedom of investment on literally cosmic scales. Here you can try to build as if the other civilizations, based on autonomous

robotics, too, with feedback and any remote control by the man himself will be fueled in any direction of development.

This fairly well-prepared base or bases - mining and manufacturing on the moon, will give a very good foundation to build cities or space ships on a huge scale - thanks to the enormous wealth of raw materials and energy (solar) on the moon.

No state bureaucracy, economic, lack of boundaries gives a chance to the incredible expansion of automation on the moon and beyond but also closer here to our earth, where the experience with the automation of the moon will be fully exploited for example to build a super mini vending machines in heart surgery, cancer. But more about that later.

Further we roll out on autonomous bases on the moon. How would work this autonomous remotely controlled these databases. As has come to a super fast pace of development of these bases. Of course, we are talking about a chain – self duplicating- reaction, just acting like splitting the atom or cloning at the same time. It is to provide the materials, energy and program the remote control stand-alone expansion.

The project MOONNOW is a very flexible design allows the freedom to determine the pace and momentum of the project. It is up to our commitment and faith depend on whether the project will remain in the realm of fiction for longer or shorter, or we will keep work out alternative routes and proposals for its implementation.

MOONNOW project proposes new solutions to drive development and mass production scale. Not like it is now - manufacture or belt manufacture more or less connected with the automation. In contrast to the belt methods used here will be the method for which the universe was created or the method of remotely guided but most autonomous Big Bang. Of course, this will be the beginning of an arduous process of submitting the first elements of automatic, organizational, transport, telecommunications, energy.

Just now the moon gives a chance for unconditional powerful development of this project. Only the super-powerful, very dynamic autonomous and remotely managed by all interested clients (using, for example, the appropriate phone applications) economy, in this case the moon gives a chance turn of civilization in a very short time.

The project MOONNOW necessarily automation & robotics coming to the fore.

This is how you start a new civilization, in this case on the moon. Robots, machines - the first of humans on earth - will begin exploration of moon, that is very rich in sources. We will only decision-makers, in which direction one way or the other exploration investments will grow. The first robots, robots base their industrial plants will reproduce (from moon sources) each other in a geometric domino effect. What can be made in a short time to an effect of a fast instant expansion of large databases, cities, space aerodromes, space shuttle plants, tourist resorts etc.

It is not only pro technology but also the design pro-social, where everyone interested in - rich or poor, infirm or healthy, unemployed, elderly or children, regardless of politics or religion or statehood will be able by means of telephony applications to support this program with simulation games, exercises construction in the increasingly numerous projects etc

-functional - but there also comes to support, through (previously described) support under the tax breaks allowances system. Everyone will get the fishing rod instead of fish to be fed to his hunger for progress in the areas, which are directly interested (cosmonautic, medicine , ecology and so on).

MOONNOW this lunar oscillator speed of development is the development platform, which must not be missed.

The project MOONNOW course start on the ground of our planet. We need to send there(moon) a "flying plate" or patched a pair of "flying plates"(we are talking about probes) with basic materials and tools for surface treatment - in the depth of the moon too, so as was at the time of Stone Age, people did the first use of the environment that they gave. With the difference that it will be now supported by the huge support of computer science, which like the beginning it will work out, preparing lunar conditions, to the automatic expansion. With time we will need more and more eager for encapsulating the moon, expanding more and more colossal plant products according to the needs of all concerned, but with a possible wider and wider support from their side. Of course, leading technology teams, as a quasi self-service restaurants, will support his work by suggestions of other stakeholders, all interested people , and by using application system – computing and mobile computing.

Already now on our Earth ground, we can experiment with for example an automatic, remotely driven plants, robots upsize plants for example in the desert Sahara or Iceland. Using every method of promotion for development of programs, simulation games for this topic project. Also, known constructions toy companies also can take part in it. Not to mention here about schools, universities, and colleges.

Only when the building on the moon reaches the ceiling, at least 1 percent of its surface and beneath the surface, we can think about a safer immediate stay of man on his or beneath its surface. Only when this lunar automatic investment oscillator is enough hot "red", we will be able to speak about the chance for successful manned space expansion and for a breakthrough in the development of robotics, automation in macro and micro scale.

MONNNOW The project was intended to indicate the effective way of expansion of the moon. However, the proposed technique outlined extensive and more super-automatics this expansion gives something much more than just the so-called the conquest of space.

This is a fully automated and remotely controlled growing to infinity oscillation growth and technological development in every field of life.

This is the most important factor that can go the progress of our civilization not I next 100 or one thousand years but in the next decade.

What does it mean fully automated - from programming production to receive such spaceship, spacecraft, space towns, and bases, robots, medical, macro, micro and nano, reception is not giving law of production process to be used by any human hand (unless robotic hand). It is this part of new style of productions – but… human hand- inhibits of growth, is the weakest link in the process of progress. That is why we need to do everything to make the weak link clear.

Using this new technology expansion and development, nothing will be as it was before today. Ships and technology from the "Star Wars" are nothing compared to the effects of earlier suggested the multiplicative speed of construction,

For many, it is too incredible to imagine a different world as we know today, and what will be completely different tomorrow and not for the tenth years. Even for me, the changes can be very drastic. Only their announcements may lead and lead today people of different reactions and shocks. A complete surprise, crying, ridicule, the recognition of the legs below the knees. Many policymakers will argue over the violation of the status quo, fear of the new.

It is a race against time to primarily save as many lives from disease and poverty, and ecological collapse of the earth. Surprisingly space, the moon has become the subject side of the main objective of the project MOONNOW but could be the main locomotive, expansions plant to achieve these humanitarian goals metaphorically and actually.

As we saw this giant outlined a very complex project MOONNOW – constructing a giant cluster of bases on the moon and beyond - requires extraordinary coordination, outstanding global technology and economy organization.

We begin this project of course on the ground (on Earth) on special dedicated parks, incubators experimental production - "robotic mankind".

In these, foundations, state agencies, companies, universities, technical and medical will jointly lead the first projects and productions in expanding copying one another objects - like robots. Only later it will spread mainly in two directions - toward the macrocosmic and microcosmic side of the super productions.

Organize, simulate, build construction objects bands incubators, which will be able configurable, develop reproduce without humans hands and use only full redundant automatic, controlled by like printers 3 d + assembly robots rules.

It will be work at the beginning like by the first cars constructing, which were slower than a horse or even a man, but we know very well who was fastest at the end - the rocket and not horse !!!

Here, the production of the "new cars" - robots or other independent organisms productions will resemble action to accelerate and improve generally production of more and better robots and rockets infinitely. Here we see the human workshop will be zero in relation to the production of accelerating oscillation. Because here already production will be dispelled, not because of the possibilities but above all - needs – economy of demand.

Let us remember for now we focus on producing products that are almost zero-occurring in the world economy today, but in the not too distant future, these new products will completely- almost 100 percent - dominate the economy. We speak here of production bases and spaceships on the moon - also due to ecology is also where it should be - for the enormity of need materials, places, and energy production would be unable on earth because of giant amounts (in many billions) mini robots or incubators robots, that would be like bone marrow cells, like army of mini robots for medical and ecologic purposes.

We're talking about a constant improvement of quality products of this super production - you know, at the start would be like with the first primitive motor cars.

Here we will have the right adjustable economics of production - he will be a realized only needs - with an infinite amount of materials and energy for micro robots on the ground and materials in space (on the moon for example). for space investment while in most autonomous program production and its multipliers, repair, and improvement of course, by the procedures and suggestions coming from people, not from machines !!! at the last stage of decisions.

This is the production of an unimaginable scale in terms of quantity and speed. It can be used (in not longer time) for example technology of supernova explosion, and in a certain time also use black holes or big bang tools for the development!

The vast extent, this will be the production of fully automatic autonomous productions and products suggested by those concerned by appropriate procedures – by for example of mobile applications- in different directions (we speak here also directions of movement in space) investment development.

Already **nothing will be the same as today. This is the end of the world! But the beginning of a new era, the era of the universe!**

Both technology sides of MOOONNOW project should support each other with help of another view of problems, chances, and experiences. Microcosmic part of project influenced by existing macro cosmos project part. But the micro cosmos part – man s body- fundamental enormously shape the core of MOONNOW self-developing robotic construction, shape the future way of present developing the project.

The core of MOONNOW project is that - robots, cluster of robots, bases, plants or towns can we cloning, reproduce like some cells or viruses or body parts of creatures like man, plant and so on. So robots will reproduce like cells self. As we know the like cells reproductions can go with very high speed and quantity – like normal in nature- and different directions of further cells robots constructing – the cells (robots) not have to be the same. That all with support with developed, regulated applications system for all interested in it.

Of course, exists the another side – it is social, economic part of the project. I don't want to make the same fault like Nobel or Einstein, where ideas very often were used not peacefully.

So with earlier suggested applications, all people should have chances of peacefully shaping her future, supporting technology or social, economy (tax) movement so this generally supports active lively creativeness whole MOONNOW system.

Starting the project would be quite easy, but the effectiveness of using outlined tools for doing this project depends on our responsibilities.

So shape our future just now !

In the end of this part of our moon odyssey, we came to a turning point of the project MOONNOW, define the mechanism of its action in the technological part.

Robots, copy machines will be self-copy like a cell of organisms by their genetic (copiers, applications programs) destination, about which we - this time- will ultimately decide.

Now effectively defined by the thesis, we can finally determine further action to move on constructing, self-perpetuating in the expansion of automated bases on the Earth - but only on start - and then on the moon and in space.

The firsts on the ground should have ecological character - for example self-extending basis of solar panels in deserts, which would then formed on the moon.

Of course, we do not forget here about very important for our health and our loved ones. Of working out like cells self-developing by self-miniaturization of mini basis, groups of mini automates, which in the near future was to the mass amount of stay in our bodies, to repair, to rebuild and in part to replace them, followed - of course according to our needs and desire - completely replace our body as you wish !

The project will be implemented in 4 parallel programs.

The space program - as the main part of the whole project – self-developing automated moon explorations,

medical program - " self-developing automatic (body) cells",

environmental program - " self-developing automatic climate shaping"

Economic program - "Automatic (citizens mobile applications) economics meet of any demand of needs of space citizens-shareholders."

Of course, all programs are global in nature.

At first, we take eco-friendly design element, and this is due to the dramatic climate change and universal experimentally nature of the project, whose first expanding plants - cells will impact directly on the further development of the so-called automatic space and medical programs too.

In short, it will be automatic expanding power plants construction for solar energy, in the desert for example. Sahara (25% the moon area) or, for example, desert Deccan - proposed a joint program of Pakistan and India. Automated factories will use solar power plants to a cleaning of air from carbon dioxide and special spraying - from produced mini drones too)powdered sand at high layers of the atmosphere in order to reduce the temperature of the Earth's climate.

This at first glance abstract expansion of the universe, which many people consider to be unnecessary waste of money, is, and will be the right engine of change in the divided, sagging ecologically world under the weight of old technology, old systems of economic and social, and already his old thoughts of native breakthrough evolution of the manufacture and production line. Developing a fully automatic, autonomous - of course in decision ultimately depends on the human indications - development of space infrastructure, among other things (and especially now) the moon and beyond, gives a chance for safe use, experimenting subsequently these technologies for environmental and medical applications.

Why only after this cosmic technological transformation can begin here on earth further technological change. Because they will not be burdened with the habits of the production by already known. "Comfortable" hackneyed(old fashioned) methods here on earth. For this, we are talking about hackneyed social relations, economic, political or even family, which make it impossible to look at whole new possibilities for technological opportunities or organizational recognition of the problem of development.

Moonnow project will be a fully open program for everyone. It is a program designed to unite all by building a peaceful empire in space and on planet Earth. The key to the success of the project Moonnow is an autonomous system mechanisms, whether large or small, which will be able to repair by themselves and also develop, multiply. It is like in a computer system to zero one nature, which we use for infinite records or calculations. Yes incubators robots will continue to grow at infinity degree in any direction depending on your needs with more or less intensive control of people. These robots or incubators or closed systems will serve us for investment, that will be on an unprecedented scale will be the only limiting the amount

These kind of closed systems will be like reproducing human cells, the reproducing virus, a cell for cloning which will be adequately controlled, producing interesting part of us and other systems at the macro and micro levels. These are our technological children, which can be driven to each other to increase the infinite atomic energy, ... energy of stars. If we are children of God, the very same way the autonomous closed systems are his grandchildren. If God or evolution gave us a brain, its development was addicted to each other from our hands and their dexterity. That our ability to effectively take actions will not only depend on the telecom computing capabilities but also on efficiency of "virus" or cell described in above autonomous mechanisms which will be more efficient than stumps current technological capabilities. These grandchildren will be those supermen- more reminding apparently coordinated groups of machines- which will be able to casually conduct mobile applications or even games..- that will be our only work casually.

We need support like project Manhattan (people free zones for testing...)- more about the Manhattan later, but on open and global peaceful scale with the support from for example mobile phone applications and other mobile services for consultation, programming, energy supply, building material or location and directions of development of the autonomous structures in extra designed for it empty spaces.

Moonnow was initially intended to be space project strictly related to invasion of the moon. But it has extended into searching and developing the use of autonomous bases, zones on Moon, the earth and also in medicine industry as well. The economics of the undertaking has spread out into pro civil, pro civilizations, developments actions. Moonnow is therefore very flexible project, open to all pro-development initiatives, particularly in technological progress of macro and micro technological enterprises.

Currently, the absolute core of this project is development of constructions of mechanisms, which is to grow and reproduce in most autonomous and remote control way for ever faster speed – like growing cluster of body cells or viruses.
This is like the core definition of the innovation.

As we know many inventions of space technology is also useful in other areas. Let's start with the main objective of the project - the Moon. It is like a big mine, the great spaceport and taste ground for exercise the autonomous development, any bureaucratic great freedom of self -replicating investments on cosmic scale. This fairly well-(automatic)prepared base or bases on the moon will give a very good foundation to build cities or space ships on a huge scale - thanks to the enormous wealth of raw materials and energy (solar) present on the moon.

Further we will roll out on autonomous bases on the moon. Of course, we are talking about a chain reaction, just acting like splitting the atom or cloning at the same time. It is to provide the materials, energy and program the remote control stand-alone expansion.

The project Moonnow is a very flexible design that allows the freedom to determine the pace and momentum of the project. It depends on our commitment and faith on whether the project will remain in the realm of fiction for longer or shorter, or will we keep working out for alternative routes and proposals for its implementation.

Moonnow project proposes new solutions to drive development and mass production scale. Not like it is now - manufacture or belt, line manufacture more or less connected with the automation. In contrast to only line methods, here it will be a method by which the universe was created - method of remotely guided autonomous Big Bang. Of course, this will be the beginning of an arduous process of submitting the first elements of automatic, organizational, transport, telecommunications and energy.

The moon gives a chance for unconditional powerful development of this project, but we can first start the moon entertainment from another full of space and solar energy like deserts zones. In this case, I suggest meetings with the courageous explorers to build the first model such as enlarging, self duplicating plants for producing solar panels for example on Deccan desert, in and for peacefully corporation of India and Pakistan.

This is how you start a new civilization, in this case on the moon. Robots, machines - the first like evolution of humans on earth - will begin exploring the moon, that is very rich in sources. We will be only the decision-makers, in which we will decide which direction one way or the other exploration investments will grow. It is not only pro technology but also the design is pro social, where everyone interested in whether rich or poor, infirm or healthy, unemployed, elderly or children, regardless of politics or religion or statehood will be able by means of telephony applications will be able to support this program with simulation games, exercises construction in the increasingly numerous projects.

It could start also by creating virtual games applications by co working with constructions toy and game companies.

The Moon as lunar oscillator speed of development - development platform, which must not be missed.

We can use every method of promotion for development of programs such as simulation games for this project. Well-known constructions toy companies can also take part in it. Not to mention here about schools, universities and colleges as these will also play a crucial role.

The project was intended to indicate the effective way of expansion of the moon. However, the proposed technique outlined extensive and more super-automatics expansion gives something much more than just the so-called conquest of the space. Using this new technology, expansion and development will happen like never before.

Space, the moon is subject side of the main objective of the project Moonnow but could also be the main locomotive, expansions plant to achieve these humanitarian goals metaphorically and actually.

This mechanism of geometrically will increase development of constructions and robots productions on the moon and other interesting areas. I try to ask for a traineeship, as it is finally turning into reality. This is primarily about the product for which requirements exceed the manual, manufacture, belt productions because here the bar is raised to the ceiling of the cosmos, both at macrocosm and microcosm level.

The idea is about a super mass production with regard to the geometric increase of speed of the powerful expansion of construction on the moon and further into space as well as the dispersal of the development of mini-robotics in medicine and environmental protection. The super speed autonomic copy productions are areas of life that should not negatively interfere in current state of the economy or ordinary activities. But in the contrary, at the prodigious extent may change the face of the economic civilization of all mankind.

The autarchic economy of the two super copy developing productions (macro and micro cosmos robots investments) is the paradigm of positive effects of the robotic enterprises experiment.

This super-multi copy production is the production of spacecraft and bases which is completely independent of the planet earth. Only the first elements such as factories, robots, mines, smelters, spaceports, urban bases will continue to reproduce from the sources of the earth and developing programs of automatically and remotely productive development. This award-copy improvement with the help of information systems will be caused by tele-autonomous and remotely controlled by the responsible teams. This will be supported by the respective territories for anyone willing to help with ICT applications and training as well as games. Because many fields or rests of cosmos will be expanded on the moon only by desire or people, who are interested in other copy super mass productions.

Similarly if separated, contemporary roaring economy will operate the automatic oscillator super-production copy on earth and it will be a mini-robotics. The separate parts of the special spaces for plants will be established for duplications and development of robots, especially in the micro version. Some would aim for full penetration of our organisms, the fight against all obstacles of health or life, which will support our cells to the degree we would like to see.

I appeal to people primarily associated with technical universities to jointly take part and also cooperate with universities so-called "hostile" countries for common development. For example, the region states India, Pakistan, Iran Bangladesh or Poland and Russia under the auspices of the universities could come together for improvement of the super macro and super micro replications production

Let us join hands with different countries, companies, foundations for our common future-present time.

Constructing giant groups of bases on the moon and beyond requires extraordinary coordination with outstanding global technology and economy organizations. Foundations, state agencies, companies, universities, both technical and medical will jointly lead the first project and production in expanding, copying objects like robots. It will later spread in two main directions - towards the macrocosmic and towards the microcosmic side of the super productions.

The human workshop will be zero in relation to the production of accelerating oscillation because production will be already dispelled, not only because of the possibilities but also demand of economy.

Both technology sides of Moonnow project should support each other with help of another view of problems, chances and experiences. Microcosmic part of project will influence the existing macro cosmos project part. But the micro cosmos part that is human body will enormously shape core of Moonnow self developing robotic construction, shape future of development of the project.

Robots(designated groups of robots) will reproduce itself like cells. As we know that cells reproduction takes place at a very high speed and quantity, the same concept will be applicable for robots constructing – the cells (robots). We will be able to regulate speed of reproduction with support of developed applications system.

With the suggested applications, all people will have chance of peacefully shaping their future by supporting technology or social, economy (tax) movement Starting the project would be quite easy, but the effectiveness of using outlined tools for this project depends on our responsibilities to shape our future already from just now.

6. Body cell-like replicating productions

Moonnow is giant and new project – about only over seven months old! -technology and social issues of planetary engineering. Self-perpetuating - programmed robotic investments (that give a new model of development for medicine and ecology too) on Moon, from materials, solar resources on the Moon It is not so easy to explain in only one or few posts, just outline of the enormous project. The different parts of the project influenced one another both in space, time and methods. We can't separately treat any parts of the undertaking - Set point economy will not be able to cope with the huge call of civilization at the present type of technological level and none the best technology will not be effective, and on the contrary will turn against us if not properly be applied to the organization, economics action.

Of course, You can take any parts of the undertaking calls for shaping by the way You, your families, your job or your school career future. You can suggest your views on constructing mechanisms of developing tools of the project.

Finally it is just project, that I propose to work out together. The project that never ends.

But You waiting for further explanations of the social and especially for the technological part of Moonnow project.

So let me do it.

Ecological part of the project would best testing start of finding better tools of work out of the perpetual mobile machinery.

Both macrocosm and microcosm level of an opportunity of using directly the selves developing constructions are not ready now to direct using them immediately effectively.

As was described before we can use deserts full of solar energy panels, as a testing area.

So foundations, schools states agencies shall start the new life, create new life. so called artificial live on the earth. First constructions, that would capable to build another the same or different constructions, machines, robots, automates, automated plants. It is of course possible!! At the start, it could be very simply plants of robots capable of building (at the start very slowly) their copies. The main goal of them shall be in quite short time build enormous self-redeveloping solar panels, industrial parks, where also would produce and developed automated constructions for powder sand and spreading this in a high level of the atmosphere to reduce the temperature of our overheated world. Of course, all the self-developed automates will finally get programs from people. The programs will be developed to speed-up the new artificial life (the self-developed robots) process like on games. Not only speed but also precisions, size and quantity of the new artificial cells (plants) on the desert.

People at start will distribute selves programs and another source for the undertaking some products, that in near future also will self-produce and distribute by the growing first mankind artificial child- that means new "person" capable to do another one

Toy and games companies can take part in the project too. From the same elements robots building another robots and plants just for playing - without specials science goals just for fun but also for developing the toys in more and more useful machines, more and more smaller, smarter or bigger and smaller, faster. Make competitions for the "games" but already not in virtually but in near future shape reality our live, shaping capacity of our body and our planet, a bit further shaping capacity of completely whole live, quality body life and expanding it infinitely. Also shaping (in not so long time!) another planets and also stars and any! other objects in the universe.

We come to a turning point of the project Moonnow which is defining the mechanism of its action in the technological part. Robots, copy machines will be able to self copy like a cell of organisms by their genetic (copiers, applications programs) destination, which will be ultimately decided by us., We will be able to finally determine further action needed to move on constructing, self-perpetuating in expansion of automated bases on the Earth.

After these cosmic technological transformation begin here on earth, further like cosmic Moonnow technological changes can be made because they will not be burdened with the habits of the production. For this we are talking about hackneyed(old fashioned) social relations, economic, political or even family, which will make it impossible to look at a whole new possibilities for technological opportunities or organizational recognition of globally problems and chances of development.

This requires courageous people who want to continue to infect this revolutionary attitude. A string of further consideration of the issues of technological change that are dependent on each other (feedback) from the physical needs, healthiness and expansion space, human and social needs will occur with planetary, stars engineering and cells engineering.

Moonnow is giant and new project –technology and social issues of planetary engineering. Self-perpetuating - programmed robotic investments that will give new model of development for medicine and ecology from materials such as solar resources on the Moon. The different parts of the project influenced one another in space, time and methods. We can't separately treat any parts of the undertaking that is set point economy will not be able to cope with the huge call of civilization at the present type of technological level and none of the best technology will be effective. On the contrary, it will turn against us if not properly applied to the organization, economics action. You can take any parts of the undertaking calls for shaping yourselves, your families, your job or your school, career future. You can suggest your views on constructing mechanisms of developing tools of the project.

Finally it is just project, that I propose to work out together. Project that will never end. But you must be waiting for further explanations of social and especially for technological part of the project. So let me do it.

Ecological part of the project would be best testing start of finding better tools of work for the perpetual mobile machinery. Both macrocosm and microcosm level of opportunity of using directly the selves developing constructions are not ready now for directly using them immediately and effectively . As was described before we can use deserts fully for solar energy panels, as testing area.

So foundations, schools states agencies shall start new live, create new live. so called artificial live on the earth. First constructions, that would be capable of build another or the same or different constructions ,machines, robots, automates, automated plants. Of course all the self developed automates will finally be programs by the people. The programs will developed to speed-up the new artificial life (the self developed robots) process like on games. Not only speed but also precisions, size and quantity of the new artificial cells (plants).

People at starting point will distribute programs and other sources for undertaking some products, so that in near future also will self produce and distribute by the growing first mankind artificial child- that means new "person" capable to do another one. Toy and games companies can take part in the project too.

From the same elements, robots building other robots and plants just for playing - without special science goals just for fun but also for developing the toys in more useful machines, smaller, smarter or bigger and faster.

We are already at dawn of history in the geometrical development, but by literally manual control, where the speed of development was dependent on the speed of propagation of people, more people more specialization. Only the method of manufacture is sometimes aided by line production. It is precisely this "impotent" point of today development, where opportunities and chances for further effective investments and productions are mediocre. This is the organizationally neglected point of speed development of technology and organization (civilization) , which we should continue to work out. But God had us to reproduce as much as possible, and it will be the same with automation. Robots, groups of robots, that will be capable of reproducing according to our guidelines, they will be additional substitutes for children (grandchildren of God, superhuman, but still obedient by our programs) that will be very helpful for development of our civilization.

I propose a project whose base is the multi application (body cell multiplication method) where construction speed is within our technological capabilities.

We are a step away from this incredible spectrum of changes that we did not even dream of. This is not fairy tale, fantasy book but the chance, that shall not lost. It is now we can build and pursue our dreams. We already have many technologies and organization experiences, that can be harnessed by us to realize at this amazing project. Body cell is a prime model for the proposed development of automation. Currently, the scale band vending robots objects will be programmed for further reproduction and development. The cell development model of production (mega production) can be a good stimulus for the development of biotechnology, medicine and environment as well.

So, paradoxically, nature becomes the basis for the "escape" from the nature by automation of robots production and development on a global scale in micro and macro scale. Project Moonnow is to be prepared in the course of solid, transformation. Of course, all primarily interested in the essence of the project which is a mega production by automatically accelerated construction development are welcomed. We will convince the world for the auto robotics development of infrastructure and the whole universe will be ours! The development of this system is the key to success of our civilizations real development. This is just the beginning of change.

MOONNOW project is a program to realize not in 10 or 100 or 1000 years but now or even yesterday!!! – Even it could already be introduced decades ago!!!

Its introduction into force of automation of self-robotic productions would accelerate- in just a few years - development of technology not only just repeatedly but a million times a billion times until infinity. Accelerate repeatedly (billions of times !!) the development of civilization as well as, among others planetary engineering, engineering of all constellations, objects in our universe knows to date.

We are already dawn of history in the geometrical development, but by literally manual control, where the speed of development literally depends on the speed of propagation of people - more people more specialization, it is because we are so "far" we went to the front since the Stone Age. Only the method of manufacture is sometimes aided by band production. And it is precisely this "impotent" point of today development, where opportunity and chances for further effective investments and productions are mediocre. This is the organizationally neglected point of speed development of technology and organization (civilization), which already we should continue to work out.

But God had us to reproduce as much as possible, and it is this same with automation, robots, that will capable of producing according to our guidelines, they will be additional substitutes for children (grandchildren of God, superhuman, but still obedient by our programs) of creator God - - high-level space – that will be very, very, very, very helpful for development of our civilization.

I proposed a project whose base is the multi application (body cell multiplication method) constructions speed is within our technological capabilities and organizational just now!!

We are a step away from this incredible spectrum of changes that do not even dreamed or dreamed of us.

These are not fairy tales, fantasy books but the chance, that shall not lose!

It is already now we can build pursue our dreams! and this project suggests various super options for automatics technological and organizational solutions.

We have already many technologies and organization experiences, that can we harness to realize this at first glance amazing project.

Body cell is a prime model for the proposed development of automation. Currently, the scale band vending robots objects, which will be delivered program - gen- for further reproduction and development. The cell development model of production (mega production) can be a good stimulus for the development of biotechnology, medicine, environment and so on as well.

So, paradoxically, nature becomes the basis for the "escape" from nature by automation of robots production and development on a global scale in micro and macro scale.

Project MOONNOW is to be prepared in the course of solid, transformation. Of course, all primarily interested in the essence of the project which is a mega production by automatically accelerated construction development are welcomed. And here we will pick properly recognition of this topic.

Convince the world for the auto robotics development of infrastructure and the whole universe will be ours!

The development of this system is the key to success of our civilizations real development.

Work out the system of self-automation robotics mega productions itself is the key to a breakthrough civilization now!

We have never surrendered!

This is just the beginning of change.

It is not only futuristic but now real possibilities for starting this project, the process.

You are witnessing live the construction of this undertaking (that is why so often my descriptions are so vague circle, not so clear) - in which you also can take part in it.

The development of super-high-speed automatic production method of reproduction is like a biological cell or meta data of sending gen (cloning) programs, which continue to steer properly prepared park robots, which would produce plants of their robots colleagues, but also like by normal cells life, sometimes some cells work will not only serve to reproduce but also for other purposes. We completely modeled by nature (of cells), which gives us an amazing opportunity to move off patterns of our civilization (ecological) tomorrow - where until today we use old manufacture methods of production - even rocket-as in the time of stone age !!! Perhaps more complicated - because milliards hands doing work now, but continue with such methods already present in nature is not capable to bear.

We cook up (literally !! -ecologic) in the earth's crust because we cannot utilize the opportunities and patterns of which (constructions, developing of body cells) just nature offers to us. We can use the simplest model of the emergence of biological life (sea of soup of amino acids and primitive great ancestors of cells).

We need to at the dawn of the formation of life to create proviso technology that we already have and never find out the hyper abilities of the automatic self-development investment.

Lacking organizational desire !! We have technological capabilities and information. We as intelligent creatures, we can send this spark the gene program to this soup of modern technological and organizational, to create the foundation to go finally to rebuild in right speed and way our development productions- out of the Stone Age!

I will tray fitting to explain – pedagogically repeatedly- point by point all the elements and concerns of the project, which does not mean that it will be boring. We say in the project Moonnow, first of all, of course about automatic mega investments on the moon by using its sources and resources. Note the word - investment, and especially the exact meaning of words automation consumption and automation investment.

So far, we have dealt with automation consumption, that is, automatic production line built by manual methods, by the stone age (manually) methods. We use this consumption automation to produce even the most interesting and advanced gadgets, but it alone in its technology performance has not moved one iota. Well, maybe it figuratively speaking, it has progressed at a rate that is a classic domino effect line- but not geometrically.

With each subsequent automatic replicating investment following plants on any amount and accelerating speed , better quality in a relatively short period of time will lead to acceleration of production to the stage for unthinkable level. This will eventually lead to surpassing power of creation to power of stars, supernovae, quasars or even the Big Bang! This means that in a relatively short time, we could attain such power in the self automatic investments that with a tiny fraction of the self investments + followed by mass scale automated chemistry needed reactions, used for military purposes could literally completely destroy our planet in a few seconds.

As I mentioned Moonnow is probably first development project covering such a broad spectrum of issues. First of all, it is a project of technology matters but its success is fully dependent on the socio-economic conditions. It has to be a program for all (without the restrictions of national, religious or what country someone comes) fully voluntary subsidized by the state or private organization. Also, the success of the correct coupling of social and economic development will depend on the proper maintenance and the level of the proposed technological change. From the perspective of smarter civilizations from outside of our solar system, we are like worms jumping on some of their adjusted site of the fire. Returning to continuations of the organization of this giant social and technological model, importance of national borders or having citizenship will disappear by the magnitude of the proposed technological changes.

We can prepare first models of the automatic creation and development of databases on the earth. This can already start from toys, games and computer simulations company's offering in education and entertainment market, models, kits, applications and different competitions. Do not forget here about all institutions, technical school or university. Do not forget about us! We also can take part in this fun educating in field of entertainment, business, learning and development. Such long-term goal of this automatic development team base is to accelerate the development of robotics in large scale investment in the space and on the ground.

At the beginning of the mass automation, we will help each other partially by remote control on the ground and in space. We will try to turn this process for all concerned as soon as possible with the help of phone applications, telecommunications on the ground and in space, among other things between moon and earth. Gradually we will be activate on a global scale by growing machine of change by this endeavor of conversion and extension of automation (as it was when the Apollo program was on a much larger scale). They will also gradually introduce developed systems of social, economic on local and global scale. In order to efficiently continue and develop this process, here technological and social processes will be deliberately influence each other on feedbacks basis.

All the programs will have on the welfare of every living creature on earth and beyond. On this account of dignity and development opportunities, in total may give us conditions to emerge from the current socio-ecological collapse, which adversely affects the effective achievement of the development goals of civilization.

We say in the project MOONNOW first of all, of course about automatic mega investments on the moon by using its sources and resources.

I note the word - investment, and an especially about exact meaning of words automation consumption and automation investment.

So far, we are dealing with automation consumption, that is, automatic production line built by manual methods, that is, by the Stone Age methods (as previously hinted at the subject). We use this consumption automation to produce even the most interesting and ever newer gadgets, but it alone in its technical performance has not moved one iota. Well, maybe it figuratively speaking, he has progressed at a rate that is a classic domino effect line - arithmetic.

What I mean in MOONNOW is an intensive use of automation to its development, its development at a rate not as much as arithmetic but geometric by quotient theoretically infinite.

What this means in practice. In practice, it looks like, with each subsequent automatic investment following plants automatically on any amount and ever accelerating speed, better quality in a relatively short period of time will lead to acceleration of production to the stage for unthinkable level- eventually lead to surpassing power of creation to power of stars, super nova's, quasars or even the Big Bang! – Creations Power of God.

In the preparation of models of automatic investment to the so-called race to the moon and beyond, seemingly incidental elements such as health or social relations are the leading factors which will shape the practical use of the project.
So if prepared models of the project will come to it, you will need more time and energy targets for auto investment for mini robotics to combat cancer and also combating diseases of old age that contributes to the death of the elderly. But to do all that, we need to start construction programs for the automation of production plants for the production of automatic robots to independently perform tasks of their investment. First of all, nature tells us to just use its resources in a natural and effective way. People have already managed to create the first living micro-creatures that can further proliferate, or cloning and stem cells proliferating, that means biological investing development and rebuilding.

We must continue to follow this step towards more controlled production of robots, assembly robots, which will also be able to reproduce, securing the supply of materials and programs. Because in terms of the real micro-robotics, we are far behind. To come to an appropriate level, we must take first care of automatically investing in space, the moon and here on earth's ecology. It is precisely here we will deal with planetary engineering in this case the formation of the Earth's climate (cooling) or in short time the climate shaping of the moon. We continue further to look at this mechanism of driving of automatic investment on earth and beyond. Automatic plants exist already, but partly remotely controlled and not self replicating. That is not fully automatic, means not fully dependent on the full programming of production and if it is already on the software, it is for short production lines. It is for the production of parts, components for consumer purposes, not for further direct investment in more efficient automation.

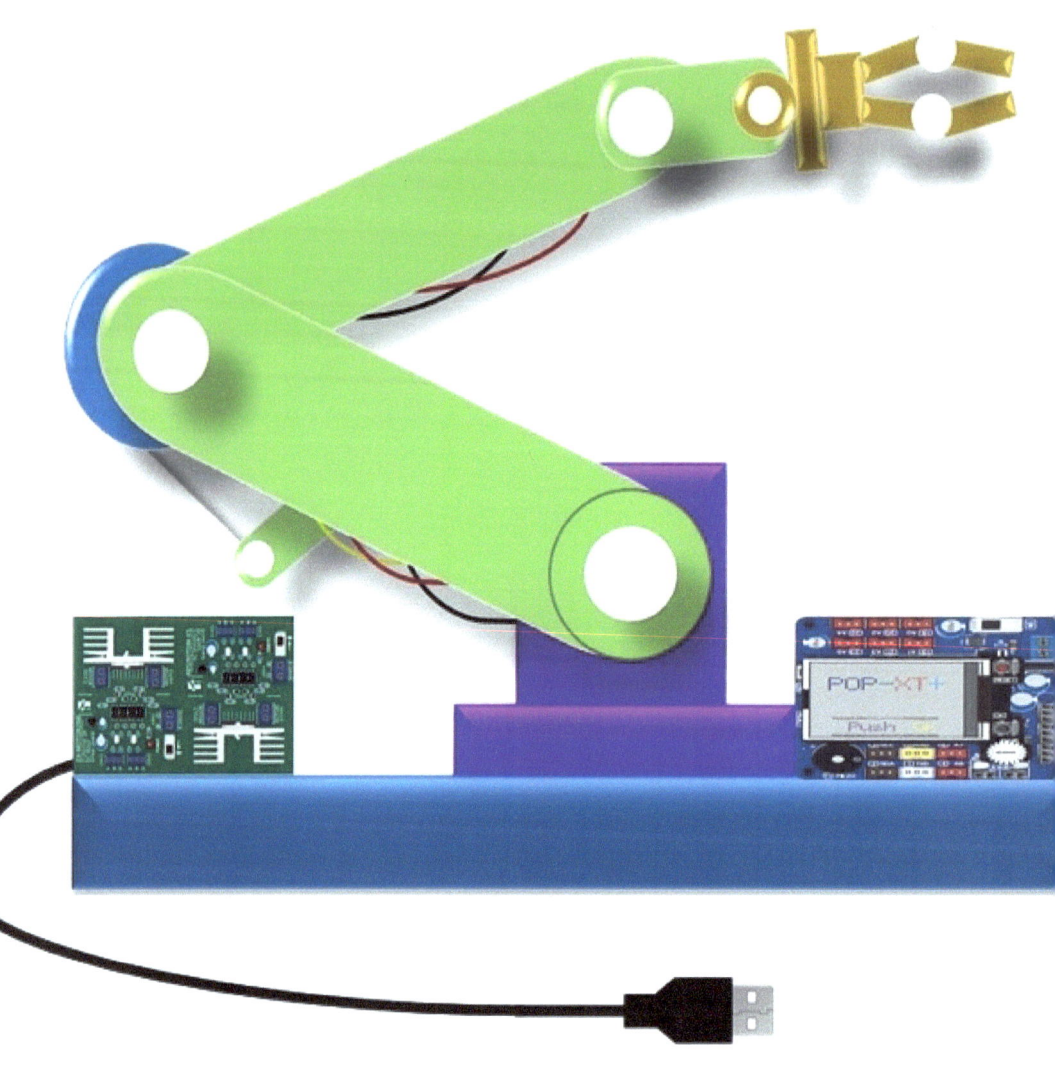

7. Real automation investment

This lack of investment in the automatic driving is the result of collective social economic habits. No one has mood to invest for the sake of further investment. But this type of investments existed often in the socialist East Bloc countries, where investments actually went full steam including astronautics in the Soviet Union, which in relation to its economic strength was much ahead as compared to USA and other countries.

That is why the moon is the lifeline for the efficient development of our civilization. Because the moon by such investments, include the extraction of materials, use of solar energy as well as nuclear. The consequences will be positive here. On earth we need to develop a more efficient robotics investments both at the macro and micro scale.

In the experimental version, we can start by smaller plants auto investing which will consist of several 3D printers. It will also include few vending machines and robots, small investment automated production. For example, in the direction of our interest such as micro submarines and drones with cameras which can evolve after continuing levels of developing investments, increase better results of these investments.

Let us not forget that the proposed automatic investments are investments in the automatics hands of a new era - not the Stone Age- by feedback with contemporary computer science will give enormous impetus (not only by just in geometrical tempo of speed!!) to the development of our civilization in all directions!

We try to figure out the essence of the accelerator self-perpetuating automatically mega investment in space - at least on the moon, or on ours ground, here in terms of mega-environmental investments. In further stages, I specifically write about medical or ultra futuristic space engineering aspects of these investments.

The essence of these investments is programmed repeatability as on a production line, with the difference that it is done by robots, in a suitably programmed system. The pre-programmed system can modulate, improve - programs and machines, robots, which directly perform program referrals task, productions, businesses, factories, production lines of machines, robots, automation, construction, and is on the moon, sea, desert, etc...

Such fully automatic (it is not a remote control, which anyway quite simply can be converted into a fully programmed !!) over the investment has never been because I did not see such a need. But now, unfortunately, it keeps up the opportunity, where not only can, but must, it is our duty for such investments starts! And it immediately!

As a full, it would be foolish to send astronauts to the moon or Mars without a powerful support with additional automatic safety bases, stations, spacecrafts on these sites and beyond.

So full of crime would be a failure of investment - selected in one of the numerous offers of the project MOONNOW- program called SAHARA, where it is proposed to boot with the construction of self–replicating automatic string manufacturing production lines of solar panels, drones and equipment for distributing powdered sand on the upper layers of the atmosphere for a general cooling Earth's atmosphere (the investments that would grow exponentially) in the desert of the Sahara, starting from states, that most willing and prepared to provide conditions for the mega project.

Before this, international organizations, state foundations supported by citizens, businesses (ecological tax breaks) would prepare an action program and the mega investment for immediate action. But self replicating system could minimize cost of it effectively.

I call for the emergence of social, scientific, political, worldwide movements to put pressure for immediate action to stabilize the climate. Then we can seriously think about further expansions such as spacecraft to the moon and beyond. Decisions as to materials, components and preparation of the first living cell accelerator of this mega ecological investment should not be delayed.

This cell system technology investments scale, even unimaginable, in a relatively short stretch of time, not only will save the entire world economy, by doing own (autarkic) mega investments on moon- not only) and so on, but actually it sprints to unprecedented scale entire economy and boost independent mega green investments and will allow the unbridled development without limiting grow without conflict that one country produces and littering more than others.

There will be no good technology without proper social facilities and vice versa, without technology, we will remain stuck in stone age. Lack of proper communication technology, telecommunications get hampered. In terms of technological progress, this will continue in the mutual dependence of the potential of human thinking and its tools. We lack the better and more modern handy tools that are needed for investments in robotics and indeed in pro civic policy. For a real explosion in technological advances, we are lacking these efficient flexible tools that needs constant development, but not those of the stone age (hands manually) , but these automatic teams of robots on Earth and beyond.

I will never resign from my plan to take a two-week trip to moon in about 10 years, but only in conditions of safety, comfort, economy and ecological affordability after the right new age technology revolution is established. Planetary engineering will take place firstly on our planet, mega self replicating investments, automated complexes for shaping the climate by constructions of mega ecological cities on deserts and oceans.

The fight against diseases such as cancer, regeneration of the human body, its functions will be taken over by the armies of micro robots. This will help in the expansion of space, where our bodies will be effectively better immunized by these micro-robots created after gaining experience of development of robotics for the purposes of planetary engineering and aerospace. This "automatic hand" may arise in these mega automated productions, backed by improved programs and applications due to better self replicating automation.

Talking about the subjugation of the self-replicating but growing structures, constructions, teams of robots, micro-robots, we not forget about concepts of self-replicating robots were also proposed at the planning of planetary engineering on the planet Venus, but was considered very futurological as there was no idea how this will effectively begin.
But now is proposing the cellular system by 3D printers and robots, remote controlled by passing to the constantly recurring programs.

We already know how this new age technology mechanism, which at the time first car (in Polish language car means **auto**), first bulb had got civilized in this vale of Earth, that in this case gives a chance for transformation striking even the biggest dreamers and visionaries.

These new age technologies are like a super virus with a super gene which is our whole wisdom and its strength is infinitely quick aggravated, mutually growing and replicating mega robotics. automation productions using 3D printers that are arranged, programmed and partly remotely controlled.

We need desire and courage to develop the programs, that are first virtual flat, later in 3D to start the revolution. We have intelligence and capabilities, we just need new hands of not only of billions of living humans but also these new automated trillions, trillions, trillions hands! It will change the face of our civilization from passive point in space until mega powerful factor positively influences active new face of human development in infinite time and space.

Intensive cosmic investments will provide better chances for fresher (technological) look at problems of our planet and its effective solutions. All we need to keep reminding ourselves that we can do it and will find substitute. It does not have to happen all at once. We need to achieve the best technology to achieve tangible chances for better troubleshooting (environmental, health) and development in space as well as in medicine. Further description of the preparations steps and stages and more concrete proposals for social and mega technological "moon" revolution will gradually take place.

Where to start the lunar revolution from?. This revolution has to be flexible. The strength of complying with changes in manufacturing and information technology. The first stage will be in the form of virtual play games, games and competitions, simulation, educational programs.

The whole process of the peaceful revolution of technology affects positive changes in the relations of economic, social, political or religious. I hinted that the universality of the peacekeeping investment marginalizes the importance of traditional citizenship, traditional national boundaries, class divisions and divisions of wealth.

In era of the mega production and indeed mega time race for protecting health of our loved ones through investments not only in the direction of the macro but also micro-cosmic spaceship, artificial cells and micro body spaceships, that can replicate itself, improving us with their own materials (worn-out cells of the human body) and external quenched materials.

8. Just for fun replicating investment

So fun, games and toys programs is a joyfully first ways and tools to happiness by "Moon revolution" – Moonnow project - first stage programs here and now, there and tomorrow, in the future, not only distant, but also the earliest. It all depends on our faith and self-denial. But enough of generalities. We need specifics hard to face the call and our ambitions. First, to practice for developing models of productions plants, because they best reflect the self-development model organism technology, which will develop in one of models of many programs of Moonnow project. Very simple models of self-replicating machines and their work have long seen. As I have written advanced technological programs perpetual motion could have use - not only on the moon - many years ago. A general description of the concept of self replication process written already long time ago, ancient times. We do not have that much to strain the brain behind this revolution, whose mission is gradual and the introduction of auto production technology investment (because it is the same replication technology already exists since time immemorial !!) to take.

It is this project Moonnow, that do possible to create comfortable conditions for carrying out the technological and social revolution. That he -the revolution- was comfortable and comprehensible for all of us each other, the need for effective communication and dialogue for our effective cooperation, the aim of which is to achieve a breakthrough on civilization scale hitherto unknown, within a few years.
But not all at once!!! You also have to try think and find up about next Moonnow proposals and steps from Your point of view.
.

How use this model of self constructing investments on Moon and other places?
How to start this process? Firstly planning of sources extracting and preparing them for assembly production. Secondly Projecting, producing, sending first completely plants and by remote control and automatically starting first steps of mainly autonomous mega productions. That could speed-up and with ever better quality. We don't have to do, replicate precise the same machines and production plants. It should work mostly autonomous like classic printing, where You offer some programs by computers, mobile

applications. We have to work out the above-mentioned process.

Some examples can help us by doing it. As I already mentioned different mobile games and constructions would be very helpful. Which theme could've most impact, depend of our and another person lively preferences. Start hyper intensive autonomous constructions investments on Moon, just for funny constructions boom, tourism, spaceships productions and science goals by most possible autarkic cost in relation for earth economy and most importantly ecology on earth, just for macro and micro robotic training and development for achieve much more ambitious goals like mega productions super spaceships, where speed would be main depend from self super speedy growing vehicle constructions.

Proposals, prototype of starting using lunar resources for push this project in practice. The same we have to work out deserts and ocean system for starting the remote mega investments by and for ecological use. I remind that finally people also can take part in it by remote control ... mobile applications and self finding ways of using resources for space, ecological and medicine useful actions. Medicine have biggest use in micro, nanotechnology also trained by virtually and educational and scientific ways.

The project Moonnow is a mega concept of civilization changes in technological and economic management in aspirations terms of space, medical, environmental and social development.

We do not need to increase costs to research and other development in this topic. Just reorganize priority expenses, for example, instead of expensive and risky trips people to the moon

or Mars (first do automatics bases and automatics mines and factories on moon and beyond), devote these costly events in the preliminary stages of the process to start in the described - by me - technologically domino effect on the moon or Mars may be, or on spaces between deserts and oceans on our planet.
And as today's programs for looking for other civilizations or exoplanets, here would be proposed direct participation of citizens in many technological projects or direct economic supports of (from taxes from every citizen separately) many programs according to the model of the project MOONNOW.
As already mentioned project MOONOW have signs of developing in the rapid autarkic - parallel - vast space economic, which will absorb, in the end, our world economy, but with a positive result for us each and every one, our civilization, our planet, having in short time impact within our planetary engineering, shape of our civilization in the universe and the formation of the universe (universe engineering) too !

As already mentioned, project MOONOW have signs of developing in the rapid autarkic - parallel vast space economic, which in the end will absorb our world economy, but with a positive result for us, our civilization, our planet, will have in short time impact within our planetary engineering, but Moon engineering too!!-
I think these self-replicating empires mega polis on the moon should achieve (by the replicating tools) in a few years about 10 percent of moon's volume !! Moon would be ultimately completely urbanized through !!! (The moon mega polis would remind the Star Wars planets shape spaceships), then we would also speak about really chances for more efficient action towards miniaturization of autonomous self-replicating robots (as it contributes biological cells) to that they could really effectively take over the roles of our body organs and cells and

further our chances of ever full health and vitality life that will need for our continued exaltation in private both earthly and cosmic life. It will shape of our civilization in the universe and the formation of the universe (universe engineering !).

What some may be bored by my economic and social digressions on topics related to astronautics and astronomy, but unfortunately it is impossible to cut off those aspects of technology because we wouldn't surely machine and cram like robots in this new era of technology. Precisely any of the revolutionary factors of this new era proposal have an imminent impact of each of us in its environmentally sustainable technological development socially and mentally.
Obviously and logically I do not fail to convey further technical details of stunning proposals revolutionary solutions.
I mentioned on formation of urban space bases (in a certain period also on earth) in blink of an eye by help of gear speed of construction growth, which would start with the self-driving up production of replicating robots, bases et cetera, on the moon using sources of raw materials and energy (solar energy).
So for a large margin, to large hooves begin mass expansion space - billions, trillions the space flights from the earth - we have to try about equally - mentioned earlier - mega ecological investments in deserts, oceans for lower the temperature above the earth.
As already mentioned instruments and methods of cell formation of large production and mine centers, solar panels, companies of space bases in space, on the moon and beyond.

And in this episode, let's look at this cell. In order to realize this complexity of the cell - virus cells constructions, operations, propagations, acquisitions of other organisms. We can its construction compared to - only one cell! - the entire infrastructure of a large multi-million metropolis, for example Tokyo metropolis.

For such level of creating such complex, effective structures at the micro or macro to achieve in short term satisfactory achievements in medicine, space, and earth, we should get started on a mass scale work out, to use, (also because our normal whole world economy and ecology would never stand such performances) system self-replicating and remotely controlled machine systems, vending machines , production facilities, databases, robots at the beginning for science and just fun space exploration case, on the moon. So self-propelling be remotely programmed and programming itself - from sources on the moon- we can form on the moon, these "cell" metropolises - and appropriately speeding and expanding the moon-metropolitan investments formed by first base and basis for future astronauts, workers, tourists or at the end of the first permanent residents.

This cell technology system of self-replicating investment gives a chance to exit (mentioned already previously) from the modern Stone Age technology (that means using today mostly, steady and constantly directly hands, welders, tweezers or syringe) that within a few years, build investments on the moon, continue intensively the exploration of the cosmos, and breakthrough in medicine in the context of parallel carried miniaturization of the cellular technology investments, or of course already shortly mentioned the mega ecological investments.

This cell system technology investments scale, even unimaginable, in a relatively short stretch of time, not only will save the entire world economy, by doing own (autarkic) mega investments on moon and so on, but actually it sprints to unprecedented scale entire economy and boost independent mega green investments and will allow the unbridled development without limiting grow without conflict that one country produces and littering more than others.

9. Taming horses and... resources

We can see almost all things in macro and micro scale. We can see cancer, exoplanets, but we need tools to gain it. Taming technology of resources and forces - by proposed Moonnow ways -we can finally achieve our apparently impossible goals in relatively very short time.
We will continue drafting of the developing model of the cell-like building replicating cosmic empires on the moon and from the mega-environmental investments on edge between deserts and oceans. Success of these investments will give a chance for efficient parallel development towards miniaturization of machines and robots (for medicine case) as well.

We would speak about real chances for more efficient action towards miniaturization of autonomous self-replicating robots (as it contributes biological cells) so that they could effectively take over the roles of our body organs and cells. It will increase further chances of full health and vitality life infinitively, that will need for our continued exaltation in private both in earthly and cosmic life.

I think these self-replicating empires mega polis on the moon should achieve (by the self replicating tools) in a few years, about 10 percent of moon's volume. Moon would be ultimately completely urbanized.
This self-replicating metropolis on 10% of Moon- that would about size of 5% of Russia or Indie size- (but with their own mines and mills of different sources, spaceports et cetera) has to fit in very broadways , then we could talk about success of our plans in areas of medical enterprises too. So in short, eternal health through the moon auto investments stimulator. Made in MOON

Gradually working according to our intentions, planets, asteroid, dark matter, stars, we should not forget about our planet, here in the first row - oceans and deserts, the ecological matters, as well as a microcosmic dimension of materials and energy of our bodies. These are so logistics extensive-area of the enterprise, that in spite of the replication system constructions activates reproduction and production, we will need masses of people into the community coordination for the help by mobile applications (this time not only virtually) of the "Made in/by Moon" enterprises. This is very important that by simulations, games or pilot-programs, educational and experiments. All people could take part in any enterprise productions (autarkic in correlations to whole "normally" earth productions) by special appropriate technology and social applications. This the key and way to come out from social and technological stone age of today.

It is time for upright posture in human and technologically evolution. In our Moonnow plans, investments, super productions, we do not have to fully reflect work of biological cells. Our products "MADEINMOON" can be different than natural, can become more efficient and better as we did historically replacing horse to racket. For example " Deep blue" computer won chess game against the best chess champion. Here in this Moonnow project, which we can also name MADE in MOON.

Horses -any energy or constructions resources- that were tamed and replaced by rockets, will become the moon, our planet, next planets and cosmic objects (parallel our smaller "horses" - cells of our bodies - to subdue), that we will tame, subdue with speed depends of added energy, matter, application programs. These new potentially macro and micro, mega-horsepower can be tamed by the self-replicating replacement development model Moonnow or similar models, that can match self-replicating machines and constructions with people's applications, phone programs and products.

The proposed space, environmental programs including medical programs can start immediately without special costs. They can aid in reducing environmental costs as well as classic economic undertaking. The advanced automation production system gives a chance for perceptible progress, for civilization's breakthrough, actual output by leaving contemporary era of stone age in terms of technologically and mentally as well. From these great yet seemingly abstract lunar investments depends our future and present of our civilization in the full sense of the word.

We must not use the hands. It's like the evolution of human development. Until man was walking on four limbs, his development manual, and thus was very intellectually limited. Today we are in same situation. If we do not get out of the stone age we will not be able to enter in the next crucial phase of our social and technological development. If we do not turn on our social transformation and technologically applications are not tamed in full automatic self replicating accelerators. This will help us to get out of this impasse of our evolution in terms of ecological, social, economic, technological, biological and decisive for our individual being as well as the entire population on a cosmic scale both literally and figuratively.

There will be no compromise on this issue. I have proposed changes for the introduction of self-replicating manufacturing and engineering grouped for the purposes of space, ecology and outright mundane - our health. Precisely, the last two words of the preceding sentence, they give strength and basis for this project to further promote and seek practical tools to meet hopes of cancer patients. Any patient and any person with desire for exciting life on earth and throughout the universe indefinitely. We already see exoplanets, we see the mechanisms of action of cells or cancer for example, but we do not work out enough for effective ways to achieve. Unfortunately we are still in alchemy age, but finally effective manual surgically tools will are proposed by Moonnow project.

As I promised, it is time to study the most anticipated aspects of Moonnow- technological breakthrough in space (micro and macro) explorations and mega productions. We have no choice but to move our technology in quite different way of development. We have to forget use of hands methods of building, constructing new machines, new tools, new towns and plants, new spaceships, new biological and other spare parts of us. So let's start final descriptions of new concept of our world, our whole universe and our whole inner body universe to exploit very intensively by everyone who are interested in it by special mobile applications programs .

Moon exploration and exploitation (huge automatic mining and building) for further intensive space explorations. Developing of pro ecological mega constructions i.e. megalopolis on edges between oceans and deserts. Race of searching and constructing of new surgical micro robotic designs for man's body. Generally it will use the system of remote controlled like "tamagotchi-care" remote controlled mobile toys, automatic constructions of intensively accelerated rebuilding.

10. Paradigm of subduing forces of nature and their replacement

The main assumption of the Moonnow project is effective, efficient, safe, low cost colonization on the moon, preceded by intensive automatic expansion.
I mentioned about using a model of self-driven automatic, autonomous technology development, expansion of production of robot's facilities bases, metropolises on the moon ,under the surface of the moon, beyond the moon i.e. on earth and in micro(nano) version in our bodies too. I mentioned that our universe will be within our reach like it is on earth. I have mentioned that our universe will be within achievable reach, that the new trend is emerging on more than a century old Einstein's and Newton's paradigms. This new proposed trend will be

paradigm of subduing forces of nature and their replacement in a better version as it was with a galloping horse 50 km/h and later rockets "galloping" many thousands of kilometers per hour. The same style of subjugation of our nature according to our needs, dreams and hopes can also be applied to journey between planets, stars and beyond and convert them and ... closer in our bodies to the pursuit of eternal health of our body.

This paradigm of a self-reliant development mechanism that replaces nature is absolutely superior to Einstein's or Newton's theory, ultimately determining the directions and opportunities for the development of our technology. Consequently, our entire civilization as well. Generally that is all what I am trying to explain is a statement of subjugation and taming of any object. The full automation of the first stage of this subjugation-venture, the expansion of the moon gives guarantees to intentions, like Einstein's theories laid foundations for atomic energy, to reshape our minds through devotions, world and universe in enormous scale.

The technological and social effectiveness of this undertaking will radiate very strong and positive effects. I mentioned about a well-known 20-year-old model of a micro robot funny under the care- tamagotchi. This tamagotchi will now be developing a mobile telecommunication application system that will give everyone ability to take part in the construction of a new parallel toy civilization, robots that will continue to lay the foundations for our spatial and qualitative expansion.

At the beginning, there will be a trifle for games, toys, more or less training or development models and specifically occurring in very extensive programs of cooperation in the development of database system, production's plants, activities of space flights on an unimaginable magnitude . These new tamagotchi "care" toy will be the architectural modular in cosmonautics in all possible aspects, based on it will be determined by architecture of expansion of new space on earth and space.

Creativity means action - not passive evolution. Do not be passive!

As I have already mentioned about evolution issues, we must remember that in process of evolution, only those will survive and have chance to develop into creative individuals who aren't passive in expectation of some "evolutionary" fate or salvation. We must courageously, actively develop the optimal conditions of effective progress - do not give up! We should not tolerate such pronouncements like " no chance, no hope" and so on. Because there is always chance and hope especially for sick and suffering and ecological situation of our planet. But now we have to go further to the most influential point of constructing the mega sustainable developing empire on the moon and beyond as well as within us. Many of people will ask, how on earth we can start such hyper mega giant project on the moon . We are not prepared so. I give the the most important answer of whole project Moonnow . This answer give chance and impetus for our entirely civilizations first on earth, in us, after this on moon and the rest of the cosmos.

We have to prepare for this moon revolutions. The proposed preparation will influence not only space industry, but entire economy, ecology, medicine, politic and our mind as well. Even small countries, big technical, medical universities, institutes of physics and chemistry, also any relevant foundations, societies, organizations and private companies could start right now to construct, build and develop in special dedicated closed, toll free, tax free, people free, international zones here on earth. It will have incubators , accelerators for self replicating and remote controlled self developing bases of robots, plants of robots and towns, as great reproductive cell. We should just not forget that most important rule of the project is " NO HANDS IN THE CLOSED ZONES ON EARTH, MOON AND BEYOND" but only fully automated (developing in size, quality and quantity and eventually remote controlled system. At starting, it could be hard, frustrating and not so effective like when first cars were constructed, they were slower than horses .

Finally, we come to the heart of practical launch of the Moonnow project. The devil is in the details. It is already known that these automatic remote-control developing and evolving lunar database systems can be duplicated here on earth in form of uninhabited islands, high-tech cyclone eye zones to force and encourage for more intensive actions toward the development of automation. This will lead to a breakthrough in space technology, ecological technology, (nano)biotechnology and nanotechnology of chemistry to maintain and improve healthy condition of all our biological organs and gradually replace them.

We need to create a simple way for the development of this lunar transition of technology according to our hopes and dreams on an unprecedented scale. It will be like a jump over evolution, an age of 1000 years ahead. We will talk about the carrot method on the international and global scales. Creation of rather permitting for selected areas with energy and material capabilities at a purely ecological level. For example- deserts can be used for the " no hands" exercises, experiments, the innovative automatic productions, structural processing, construction of mega macro and micro. It would be like training a monkey to use box or stick to catch bananas. It is enough to designate, define the international legal framework relief of any possible tax, any bureaucracy, any tax and customs.

I would like to remind you that these territories of new age technology will be autarkic, like first computer programs or computers, first cars that have contributed much to further economic or civilization.

We come to realization stage of visions and concepts presented in the Moonnow project. It is great art to present new visions of a better world, a better future for present world, but even more important is their specific preparation and realization. There is nothing to lose here, and much to win for the launch of this socio-economic technological domino effect. There is a need for concrete cooperation with people, organizations, institutions, universities. Specific dialogue, exchange of proposals and comments is needed. That is why you need to be more active. I will continue to point out more precise, specific and practice developments on this topic.

I mentioned the moon as a training ground for boundless experiences with development of space bases, cities, and mines, power installations (solar panels) and other installations and production that would largely allow for general development of macro as well as micro (nano) technology. I also mentioned reversed version of the Manhattan project, that is, the proposal to use desolate desert areas as opposed to chemical chain reactions. Atomic destructive would be used contrary to create lunar development processes described above with additional support of solar panels (in Iceland it could be geothermal). Of course as long as in these isolated zones, the process of self-propelled production and self-sustaining infrastructure expansion, there will still not be sufficient developed processes for handling chemical reactions to independently produce materials needed for more efficient production and mass-scale production.

Resurgence of technology for everyone.
This is the most important part and also the last one. The mission of presenting the project initially aimed at automatic self-replicating expansion of the moon is almost fulfilled. The mission of showing a specific path, a concrete tool for the resurgence of humanity on a literal and metaphorical scale. It will be achieved by geometrical pace of speed for cell-like self-replicating automation in cosmonautic and other technological development departments of humanity.

11. Peoples free zones – small "moon" zones for everybody

In the western world celebrate Easter it is the feast of the resurrection of Jesus Christ, the resurrection, where the egg symbolizes this event. This is just an egg, the rebirth of a new life will be a great element and a symbol that accurately explains idea of conception of MOONNO project. The proposed zones as incubators or mass-market automation accelerators to occupy the moon or other cosmic objects can be found also on relatively large deserted areas on earth, that are rich in at least ecological energy sources, that can be found in concrete preparations for unimaginable scale in human history. The so-called experimental studios, polygons – smaller(much cheaper) scale zones could be also in use so that even schools, single persons or foundations, agencies, companies could participate in this!!!

The basic principles of running these mini zones of programmed and remotely driven, training terrains for geometrically multiplication, duplication of robots are:

all materials or energies (from outside) can lead to boundary of these zones, if they have not sufficient resources of them,
there must be enough room for self-replicating multiplication (of course ever faster, more and better quality) for next-generation same set-plants of robots by help of for example 3D printers, but.. of course, by no help of any person and his hands, that wouldn't have right to be in these zones - so not allowed to put in the zones another machines- eventually components. Only the first set may start the process of "the automatic generic evolution" in the zones, by help of development for programming and for remote control.

This is the chance for all of us including every country or institution to enter into newly reborn human development opportunities **movement** on scale beyond any of our imaginations in technological, social, economic, cosmic and human dimension of mankind.
You in your environment have got the drafts and tools to push our civilization step further. Now it's your turn to act! It is time to take first step for constructing details because the mission of presenting the project initially aimed at automatic self-replicating expansion on the moon is almost fulfilled. The mission of showing a specific path, a concrete tool for the resurgence of humanity on a literal and metaphorical scale. The "moon" construction, first on

earth should start from organizational matters and it is about in the extra ordinary situation of the extraordinary construction foundations for waking up, stimulating and spreading awareness to scientists, technologists and citizens. We need to create organizational foundations for civilization jump that more precisely could fit our individual and collective needs, dreams and potential enormous opportunities. All this with exclusion of religious, national barriers and political parties.

A global "environmentally, cosmic-friendly" foundation based on a voluntary partnership of scientists, technologists, citizens - something like the doctors without borders would become a model of public relations for now and tomorrow, here on earth and in space. It would take us in next future from today's systems (governments, economy) to a smooth degree. So you in your environment and with your social environment can start initial movement, Moonnnow's offers and needs to take final practical steps. Be not passive, be constructive.

Moonnow is about creating peaceful global and local organizational and technological conditions for hyper active participation in local, global, cosmic transformations in full sense of the word. What is the purpose of the Moonnow project? Are super cheap passenger flights to the moon and beyond? Yes, but it is just an epiphany. It is a string that draws attention to the proper teapot and development path. It is a shift from evolutionary social economic and political stagnation to goals above all to move to a higher level in development of plans for visions of constructions that will completely shape us and our civilization of our surroundings technology on a macro and micro scale of our cosmos in literally and immediate mode.

We have full potential development talents, opportunities to overstep the impasse of primitive social and political economic relations with the help from appropriate organizational and technological change. The courageous decisions in building the foundations of technological hyper-technological as well as engaging in this strengthening of proper socio-economic bases will give chance for true developmental change in the broadest sense of the word.

It is not about replacing the theory of evolution with the theory of creativity, but by leveraging the achievements of the evolution of technological evolution into further irrepressible development, which can be characterized by creativity and hard tangible evidence of the existence of God as the omnipotent creator. We can create things, creatures and worlds with a flick! Proposed technological solutions, which will be described in details in this mega construction, have full grounds for creative evolution for the creation of cities! There will be no time wasted and we should not waste our chances and talents to make world better place because creation and evolution together give tools, consent and ordinance to make our world more human, divine and merciful. We take a big step to disclose the construction details of the Moonnow project.

12. Amanhattan

In the next step of the suggested proposals of this mega social-technological project, I will explain concepts of the so-called reversed version of the Manhattan project. "Manhattan" project as mega destructive, mega constructive speed and accelerator project draft for further analysis.
We will use more power to reconstruct our life process but without destruction. Just opposite of project Manhattan would be best model to describe the proposal of Moonnow.

Moonnow construction will change us and we in new way will influence its development in feedback accelerators. It is like racing with diseases, racing with(against) death, racing with stars and light. It is racing with our steady narrow-minded, old attitude to life, family, culture, religion, nationality, political, economy, literature, medicine, health. We have to reconstruct whole model of our life and future. We can behold tradition but from another point of view.

We have nothing to lose by achieving our goals aiming shift of our development. We need to use our potential of materials and technological organization resources.

As I have very often said, my articles are calls for wake up from unfaithful attitude of our place and missions to make more human on behalf of God. Of course they are not just empty calls but suggestions and proposals for your local and gradually global activities and initiatives. It is giant project and because of this we need space energy on earth, moon and beyond to effectively achieve our goals of shaping the body of universe including body healthiness, harmony and dignity of everybody. We speak not only about economy, technology, political, ecology, medicine, astronautic separately or mixed one another. This is a completely new way of rescuing and developing our life in very broad way by space engineering.

But we need free platforms to start this hyper fast process. Yes this is like war against bad things that blocks ignition of new big bang evolution and reconstruction that can positively fulfill our mankind voyage described as alder evangelism (fulfill and extending visions of really developing mankind in universe) and dreams. We have to use most powerful tools to do it . Any institution – as foundation- can start independently and separately this cosmic process of transformation which would be capable to start construction at even bigger remote controlled industrial park, reservation park where self copying machine and tools would play the main role.

It is extremely important to point out as often as possible that main core of this hyper new concept is that only first robots and automates will be built by people. The first robots like first new remote controlled people have to develop and multiply from materials by fast programming the next generations of robots capable of self coping and production of new rockets and so on, with help of mobiles applications.

It would remind you of project Manhattan, where scientists and engineers worked full confidential in deserts in restricted areas to build up bombs to faith evil sources of second world war. This time also we would faith- starting also from deserts- evil sources of diseases and by developing to use right resources for growing our development expansion in a blink of an eye.
Inverted Manhattan project means that this time everyone can take part in shaping acceleration of positive technological development with organizational media support - telephony applications by supporting global and local social contribution proxies. As in the Manhattan project, which was spotted for war purposes, here it will be for global technological initiatives that will completely change the face of sadly divided earth by militarily potent warfare intersections into hyper-technological self-articulated urban space-based medics, especially in the direction

of creating controlled mechanisms to speed up the pace with self built nano robots technology.

Here it will be used energy for non-massive design, but a construction that will continue to increase affinity of geothermal energy resources, extraterrestrial solar and stars – in time. Energy on mass scales to supply resources of the Earth and beyond like " big bang constructive explosion" - not like the atomic bomb destructions manner.

Just in contrast to Manhattan's project, it is a humanitarian project despite the proposed advanced automatic pro in every field of life, especially in medicine, astronautics and ecology. The global goal (across borders and other artificial interfaces to the development of divisions) is to address problems and opportunities for development. Giving everybody chance for organizational, political, technological swing to the most advanced stage of human evolution and revolution.

The Manhattan project was a project of the elite , who at the same time made the technologically successful arms race in the "peaceful" annihilation behind the backs of the masses, behind our backs, and "in our name" may be as unwary as the Einstein or Nobel said. In my initiates and inspirations, you have a special inclination to avoid these mistakes.

Regarding the gradual disclosure of project details, it is only the beginning of this process (plan) that you will at some point be invited to participate directly. There is no way of classical description or so-called ordinary language to describe all the interdependent nuances of this megaproject.

We continue step by step construction analysis of the new age hyper advanced powerful physically, constructional, chemically technology as MOONOW project, which due to its similarities and contradictions, has been drawn as the manhattan project - (referring to the project of using nuclear energy to build the first atomic bombs)

Similarities are there, but are not limited to radicalism of problems solutions. Here we speak of technological radicalism to the problems and opportunities of political and socialist views on counter-primitive relations by nationalist thinking, as well as a military adventuring technological and organizational solutions.

Also, safety experiments should be carried out in our project for effective experiment of this adventurer explosion, i.e. structural, production, construction to achieve design goals in a moment in comparison with most advanced adaptive or structural processes. Just already a few such(auto replicating productions) centers on earth can replace - in a fraction of the time – could overtake whole world production and development.
But next difference of this manhattan is its technological abundance, regardless of origin, place and age of all inhabitants of the planet. So everyone from now, individually or in a group can (by mobile tools) start this healthy contamination –replicating- technology by building from building constructional, energetic, material components such as building blocks and other subcomponents to build just for fun or normally houses, spaceships et cetera.
I provide the hyper project reminding bypassing system, generally speaking parallel to so called to normal life of today's economy,

society and normal service and technology, so that give guarantee effectiveness of building new age technology system without dangerous interfering.

13. We are only little step from being masters of whole life and universe

So we coming to the last summary chapter of the cosmic bold draft book of a human but literally" hands-free" transformation of oneself and universe.
 The book could be used for fiction or impulse to real action. We can see stars or cancer but we can't gain it. But showed possibilities lead us to gain it almost just now. Like by

program Apollo we have to use heavy organizationally power and ... powerful social movement- in contrast to Manhattan project- (At Apollo program there were hundreds of thousands of professionals and volunteers) to launch this project, but internationally and almost for everyone actively not passively, from constructions games and toys, any school, companies, private persons mostly by using mobile applications programs.

Desert, Moon like on Manhattan project is the best place to start the building constructions productions accelerators that would mainly not use for extract energy power but hyper multiplied program productions. Power also but with a heavy manipulated system of technology that will be used also to developed speeding any chemical reactions needed for us – heavy water and normal water too from everything, and specifically developing technology that helps people to be functional, effective in not always appropriate conditions. Production of replicating robots for developing gradually ever faster and better robots and by the way other productions but for God sake never with help hands of people!!!!

Only initially by hands build robots - not forget, that initially robots will more remind cluster of robots- will be put in peoples free zones, where they robots will take their own two-way (macro and micro/nano) evolution, but this time with help of developing remote controlling applications of another robots , where mainly key of the proposal in my projected revolution is programmed autonomic developing with little influences of remote controls programs.

This way we can empower right speed of technology development in directions of our all wishes and dreams. Lunching technology on an unimaginable scale.

Anyway, we are on edge of the revolutionary outburst. We electronically almost ready to go out into space from our today place, out of our cave of being on the earth and in mind. We need just harness our stone -steady like - age technology, mostly manual productions in right tempo of constructing and building.

We have to build our first child of mankind not duplicating first God child, not already not perfect (spoiled)child of God, but own child or grandchild of God – we are (have to)already enough mature for it- what it means is the birth of the first grandson of a god, as artificially man robot that can self replicating and developing with genes from us from children of god genes of programs, that we capable, can produce by the way steering with programs and robots like with our children by raising and educating.

There would be harsh environment for it apparently like for first live beings on the earth, but we can do it better and better teaching our artificially children for search and use any environment in peoples free zones like moon and any space and objects, that can give opportunities for effectively independently of hands, programming a bit remote controlling development. developing .

By further developing of the self-automation, we developing chemistry but also after that physic of anybody celestial body that further could influences physically world and universe in any dimension so shaping literally whole world and universe gradually more and more according to our wishes and dreams.

We are only little step from being masters of whole life and universe

But somebody can say that we haven't still robots so effective independently developing machines.

Yes, it is true that is why we have to do it!!!

The first step of that would be building robots parks-with an effective supply of materials, components, and energy- like body cells from small or little bigger blocs machines blocs and 3d printers, that could by programming continuing the multiplication just for research and fun. Even any toy companies can start the process of the domino effect, just now with help of any heavy support of mobile applications for/of everyone, a child too!!

I described it all would remind first body life cell multiplication in first extra for this developing environment programming, technologically and organizationally economically.

We can start free zones of tax and any economic state duty, that allow for any enterprises start without any state money, experiments for hyper automation productions. So no one could say we have no money for it. The free economically free people zones can do it can push further our project involuntarily automatically.

But somebody can say that we haven't still robots so effective independently developing machines.

Yes it is true that is why we have to do it!!!

Production of self-replicating robots for developing gradually ever faster end better robots and by the way another productions but for God sake never with help hands of people!!!!

Only initially by hands build robots - not forget, that initially robots will more remind cluster of robots- will be put in peoples free zones, where they robots will take own to ways (macro and micro/nano) evolution, but this time with helps of developing remote controlling applications of another robots. First step of that would be building robots parks-with an effective supply of materials, components and energy- like body cells from small or little bigger block- parts assembled machines and 3d printers, that could by programming continuing the multiplication just for research and fun . Even any toy companies can start the process of the geometrically

domino effect, just now with help of any heavy support of mobile applications for/of everyone, child too!!
I described it all would remind first body life cell multiplication in first extra for this developing environment programming, technologically and organizationally economically.
Any way we are on edge of the revolutionary outburst - Lunching productions technology on unimaginable scale .We electronically almost ready to go out into space from our today place, out of our cave of being on the earth and in mind. We need just harness our stone -steady like - age technology(mostly manual) productions in right above described replicating automatically tempo of constructing and building.

www.ingramcontent.com/pod-product-compliance
Lightning Source LLC
Chambersburg PA
CBHW040220220526
45473CB00001B/63